THE PETERBRO TRACTOR

MANUFACTURED BY
Peter Brotherhood, Limited,
PETERBOROUGH.

Smithfield
Show
Stand No. **2**

**BRONZE
MEDAL,**
LINCOLN, 1920.

For use with Howard Ploughs, Cultivators, Knapp Drills, Garrett Threshers, and other Farm Implements.

AGRICULTURAL & GENERAL ENGINEERS, LTD.
CENTRAL HOUSE, KINGSWAY, LONDON, W.C.2.
ASSOCIATING :

Aveling & Porter, Ltd. (Rochester).
Barford & Perkins, Ltd. (Peterborough).
E. H. Bentall & Co., Ltd. (Heybridge).
Blackstone & Co., Ltd. (Stamford).
Peter Brotherhood, Ltd. (Peterborough).
Charles Burrell & Sons, Ltd. (Thetford).
Burrell's Hiring Co., Ltd. (Thetford).

Clarke's Crank and Forge Co., Ltd. (Lincoln).
Davey, Paxman & Co., Ltd. (Colchester).
Richard Garrett & Sons, Ltd. (Leiston).
James & Fredk. Howard, Ltd. (Bedford).
L. R. Knapp & Co., Ltd. (Clanfield).
E. R. & F. Turner, Ltd. (Ipswich).
A. G. E. Electric Motors, Ltd. (Stowmarket).

Stand No. **12** SMITHFIELD SHOW.
International Tractor Trials, held at Lincoln, October, 1920.

THE
BRITISH WALLIS
BRITAIN'S FOREMOST
TRACTOR

was awarded by the Royal Agricultural Society of England,

**FIRST
PRIZE** ❧ **GOLD
MEDAL**

in the Three-Furrow Class.

**PRICE
£525**

*Present
delivery
seven days
from date
of order.*

Manufactured by
**RUSTON &
HORNSBY,
Ltd.,
Lincoln.**

For further particulars apply :—

78 - 82,
BROMPTON RD.,
LONDON,
S.W.3.

Managing Director

Cynddylan on a Tractor

R.S. Thomas

Ah, you should see Cynddylan on a tractor.
Gone the old look that yoked him to the soil;
He's a new man now, part of the machine,
His nerves of metal and his blood oil.
The clutch curses, but the gears obey
His least bidding, and lo, he's away
Out of the farmyard, scattering hens.
Riding to work now as a great man should,
He is the knight at arms breaking the fields'
Mirror of silence, emptying the wood
Of foxes and squirrels and bright jays.
The sun comes over the tall trees
Kindling all the hedges, but not for him
Who runs his engine on a different fuel.
And all the birds are singing, bills wide in vain,
As Cynddylan passes proudly up the lane.

TRACTORS AT WORK

A Pictorial Review
Volume Two

STUART GIBBARD

FARMING
PRESS

First published 1995

Copyright © Stuart Gibbard 1995

All rights reserved. No part of this
publication may be reproduced, stored in
a retrieval system, or transmitted, in any
form or by any means, electronic, mechanical,
photocopying, recording or otherwise, without
prior permission of Farming Press Books & Videos.

ISBN 0 85236 316 8

A catalogue record for this book is available
from the British Library

The poem on the first page is published by permission from R.S. Thomas,
Collected Poems 1945-1990, published by J.M. Dent.

Published by Farming Press Books
Miller Freeman Professional Ltd
Wharfedale Road, Ipswich IP1 4LG, United Kingdom

Distributed in North America
by Diamond Farm Enterprises
Box 537, Alexandria Bay, NY 13607, USA

Cover design by Andrew Thistlethwaite
Typeset by Winsor Clarke, Ipswich
Printed and bound in Great Britain by
Butler & Tanner Ltd, Frome and London

Preface

The decision to write a second volume of *Tractors at Work* was really taken out of my hands. When I started to write the original book, I had no idea what a maelstrom of interest and memories I was stirring up. I was caught unprepared by the amount of feedback I was to receive after the book was published, and was delighted by its warm reception.

The idea behind *Tractors at Work* was a selfish desire to create the book that I had always wanted and felt should have been written — a book that, through period photographs, actually shows the tractor working as it was, warts and all. I was not going to discriminate on age limits — to me any tractor is interesting whether it be built in 1905, 1945 or 1995, and all are equally important in their own context.

From all the comments, information and offers of photographs that flooded in following publication, I realised that many people were of the same mind and that *Volume Two* was a *fait accompli*.

Listening to readers' reactions to the book, I became aware that the photographs showed far more than just the development of the tractor — each picture captured a point in time; a part of agricultural history that went beyond the impact of the tractor and evoked many poignant memories of a long-forgotten past. Not only had the tractors changed, but the farming scene had altered forever.

I soon realised that the tractors were important — but without the man on the seat, they were nothing. It is with this in mind that I dedicate the introductory chapter to the tractor drivers — the men who spent their lives working with tractors.

Tractors were used for many purposes other than agriculture, and this time I have tried to choose photographs which illustrate the machines in a wide variety of applications; otherwise the format remains unchanged.

Once again, this book would not have been possible without the many people and companies who have kindly searched out and loaned photographs, or allowed permission for photographs to be used — in particular, Jeremy Coleclough and Gerry Wright at Opico, John Briscoe at Massey Ferguson, John Hewitt at New Holland, and Steve Mitchell of Pharo Communications for John Deere photographs.

Special thanks go to Clive Brown, Editor of the *Lincolnshire Free Press* and *Spalding Guardian* for permission to use photographs from those two newspapers, and to Tim Wilson for looking out the negatives; also to John Blackbeard, Editor of *Arable Farming,* and Graeme Kirk, Editor of *Dairy Farmer,* for allowing me to sort and use photographs from their archives. I am also greatly indebted to Eric Sixsmith, Peter Longfoot, James Coward, Geoffrey Gilbert and Derek Hackett for assisting me with invaluable background information, and my wife, Sue, for once again patiently helping me organise the layout of the book.

Finally, I would like to thank all the individuals who have sorted out precious photographs from family albums for inclusion in the book. Keep them coming: *Volume Three* is always on the horizon.

Stuart Gibbard

Picture credits

Colour plates in **bold** type

Peter Anderson: 5, 47, 100, 101.
Stan Anderson: **3, 4.**
Arable Farming: 143, 148, 149, 156.
David Bate Archives: 28, 32.
British Sugar: 74, 75, 77, 114, 132.
Stephen Burtt: 14, 55, 65.
Arthur Carter: 52, 63, 70, 71, 78, 80, 174.
J. I. Case Europe Ltd: 108, 152, 153.
Mr Chappell: 26.
Willie Chatterton: 42.
Chris Chouler Collection: 19.
Dairy Farmer: 83, 84, 99, 105, 106, 147, 159, 164.
Ven Dodge: 33, 34, 56, 57, 58, 155.
Ernest Doe & Sons Ltd: 134.
Fred T. Dyer: 51.
JCB Landpower: **13.**
John Deere Ltd: 173, **16.**
Lilian Ream Exhibition Gallery: 38, 39, 40, 44, 45, 46, 50, 67, 79, 81, 87, 102, 120, 122, 123, 129.
Lincolnshire Free Press: 62, 64, 72, 76, 86, 89, 103, 111, 112, 115, 116, 124, 126, 127, 128, 130, 133, 135, 136, 137, 138, 139, 140, 145, 146.
Massey Ferguson (UK) Ltd: Frontispiece, 93, 161, 162, 163, 171, **15, 17.**
Alan Mole: 1, 2, 3, 73, 94.
New Holland Ltd: 49, 169, 170, **14.**
Opico Ltd: 142, 144, 150, 151, **5, 6, 7, 8, 9, 11.**
Quadrant Picture Library: 8, 9, 12, 13, 16, 17.
F. Rowe: 29, 48.
Silsoe Research: 35, 36, 53, 59, 61, 68, 69, 82.
Eric Sixsmith: 6, 7, 10, 18, 23, 24, 25, 43, 66, 95, 96, 125, **2.**
Mr Smalley: 107.
Frank Smith: **1.**
Mervyn Spokes: 60.
Brent Usher: 85, 117.
Robin Wheeldon: 21, 30, 31.

Introduction chapter photographs: Willie Chatterton, Lilian Ream, Stephen Burtt, *Dairy Farmer,* John Deere.

The author would also like to acknowledge the following people who have located or loaned copies of photographs: Colin Boor, Ted Chamberlain, Ken Hatter, P. & S. M. Johnson, Brian Leake, David Lockhart, Rex Thorpe, Rob Thorpe, Mrs Waters.

Bibliography

Baldwin, Nick, *Kaleidoscope of Farm Tractors,* Old Motor Magazine 1977.

Bell, Brian, *Fifty Years of Farm Machinery,* Farming Press 1993.

Culpin, Claude, *Farm Machinery — Tenth Edition,* Granada 1981.

Hanley, J. A. (Editor), *Progressive Farming,* Caxton Publishing 1951.

Hine, H. J., *Tractors on the Farm,* Farmer & Stockbreeder 1942.

Irwin, Michael D. J., *The Nuffield Album,* Allan T. Condie 1993.

Longfoot, Peter J., *Caterpillar Tractors 1926-1959,* P. J. Longfoot 1993.

Royal Agricultural Society of England, *World Agricultural Tractor Trials — Official Report,* 1930.

Sherwood, George, *The Farm Tractor Handbook,* Iliffe & Sons 1919.

Stirton, John (Editor), *Transactions of the Highland and Agricultural Society of Scotland,* William Blackwood and Sons 1918.

Walker, Allen, *Inside Story,* David Brown Tractors 1977.

Williams, Michael, *Great Tractors,* Cassell & Co. 1982.

Williams, Michael, *Classic Farm Tractors,* Cassell & Co. 1984.

Wright, Philip A., *Old Farm Tractors,* David & Charles 1962.

Introduction: Working with Tractors

The degree of comfort and sophistication offered by today's technologically advanced tractors belies the harsh realities of the life of a tractor driver in the not too distant past.

The modern driver, used to a tractor that instantly bursts into life at the turn of a key, would be horrified by the complicated starting procedure facing the operators of the pioneer machines built eighty or more years ago.

Preparations for starting these early tractors included filling and manually priming the mechanical oiler that lubricated the cylinder, crankshaft and piston; setting the needle valve to adjust the flow of fuel; priming the fuel pump; opening the compression taps; retarding the ignition lever, and finally pouring a few drops of petrol directly into each cylinder. Naturally the starting sequence would vary from make to make, but whatever the model, it was still a time-consuming task.

The Farm Tractor Handbook, published in 1919, gives elaborate details of the preparations involved, concluding with the optimistic statement, 'One or two smart turns of the starting handle should now suffice to start the engine.' Of course, this does not take into account cold mornings, damp magnetos and oiled-up plugs — swinging the handle until blisters formed and tempers frayed — and the all too frequent backfire that could result in a broken arm.

With most tractors in Britain fuelled on paraffin or TVO (Tractor Vapourising Oil), the driver had to contend with starting on petrol and changing over to the paraffin distillate after the engine was warm. Plumes of white exhaust smoke often indicated a driver's over-eagerness for a too soon changeover when trying to conserve expensive petrol. Wartime petrol rationing only encouraged the habit, and tell-tale streams of white vapour emitted from the exhaust pipes of many farmers' cars demonstrated that vapourising oil was never in too short supply.

The early tractor driver was afforded little comfort — a padded seat, the heat of the engine and the luxury of a pipe.

Electric starting on tractors did not become common-place until the late 1940s, its use further encouraged by more widespread developments in the utilisation of the diesel engine, which was difficult to start by hand.

The early diesel engines, with few exceptions, were notoriously bad starters requiring complicated cold-start devices, blow-lamps, lighted tapers or heater-plug circuits to persuade them to stir grudgingly from their slumber with a clatter and knock that would send the rats scurrying from the barn. Frosty winter mornings would often lead to desperate measures — a flaming diesel-soaked rag pushed into the air-intake, a tow around the village with another tractor or, as a last resort, a squirt of ether from the 'Easy Start' can down the air-breather.

Winter presented the tractor driver with other problems. Anti-freeze mixtures, usually of glycerine and alcohol, were not always widely available and were expensive. If they were not regularly checked, evaporation of the alcohol often left the mixture less than effective, and many early tractor manufacturers would not recommend their use. In frosty weather, most tractor

Above left
The driver of this Case 'C' could look forward to many hours sitting on a cold pan seat.

1940s tractor drivers line up with their machines — two Allis 'M' crawlers, an International TD6 and two John Deere 'B's on a Norfolk farm. Note the early weather cabs.

drivers drained the water from their cooling systems at the end of a day's work. In the morning, pans of hot water would be found boiling on the stove, not just to make up the can of tea which would be drunk cold at lunch time, but for refilling the tractor radiator as a warm engine was easier to start.

Crawler drivers knew better than to leave their machines standing out in a wet field on a frosty night without first driving them up on two wooden sleepers, or the morning would find the tracks immovably frozen to the ground.

Although Harry Ferguson's little grey TE20 is often credited with revolutionising tractor development, the new Fordson Diesel Major, introduced in 1951, must be acknowledged for improving the lot of the tractor driver.

Lincolnshire, September 1941 — the driver of this Fordson receives his instructions for the day from the farm foreman before taking a load of machinery to the next farm.

A 1941 advertisement for a weather cab to fit the David Brown VAK1.

ALL WEATHER
CULTIVATION
with the DAVID BROWN
general purpose tractor

The folding, car-type canopy illustrated can be fitted in a short time by any handyman and gives generous protection even in the most boisterous weather. Delivery from stock at Meltham. Retail price £7.10.

DAVID BROWN
TRACTORS LTD.   MELTHAM   Nr. HUDDERSFIELD

Safety frames and cabs, introduced from 1970, were designed to save lives in over-turning accidents.

Here at last was a tractor that, with the press of the cold-start button and a quick push down of the starter-lever, would instantaneously burst into life and go straight to work without further ado. This uncomplicated and dependable engine would slog away all day without any more attention. Few tractors since have started as easily.

Although the Diesel Major offered many advantages for the driver, comfort was not one of them. Even a seat cushion was an optional extra. Today's tractor driver takes for granted the level of refinement afforded by modern tractor cabs — but these are only a relatively recent development.

The early tractor driver would not receive or expect any protection from the elements. The horseman had always worked in all weathers, and the tractor operator would not want to be treated any differently. His only luxury would be a hessian corn sack to insulate his back-side from the cold metal of the pan seat. This sack also had other purposes — it was hung over the tractor radiator to encourage a faster warm-up for a quick turn-over to TVO, and on a wet day it would be placed over the driver's knees.

Little other thought was given to weather protection for the operator. H J Hine, in his wartime book, *Tractors on the Farm,* recommends, 'My own experience is that the best clothes to wear in wet days on the tractor are the waterproof overall suits sold for motor cyclists.' Many drivers, however, preferred an ex-army greatcoat.

The weather cab began to make its appearance during the Second World War when it became important to keep the tractor working in as many weathers as possible. These early cabs were crude affairs, often fabricated from canvas or wood, before improved sheet metal cabs came on the market in the 1950s and 1960s.

Few further improvements were made to the weather cab before the launch of the safety cab in 1970, designed to protect the operator from being crushed in the event of the tractor over-turning. This cab, in itself, brought its own problems — its rigid metal cladding resonated with the sound and vibration of the engine, and long days on the tractor seat left some drivers with hearing problems, leading to the introduction of the quiet cab in 1976.

Not only was the driver now protected from noise, he was also insulated from the warmth of the engine — the new quiet cabs were like iceboxes in winter and larger windows for improved visibility only served to trap the sun's heat in summer. It was not long before heaters and blowers became standard equipment with air-conditioning as an optional extra. Radios were fitted to make up for the missing decibels. Further developments in the 1980s led to power steering becoming standard, and the introduction of electronic controls.

The technological advances of the 1990s have seen the modern tractor cab become a state-of-the-art working environment with computer-monitored instrumentation systems and the programmable controls at the driver's finger-tips. The new John Deere 8000 Series tractors, the first in history to have their design concept patented, boast CommandView cabs with all the major tractor functions incorporated in a revolutionary new CommandArm armrest control module, which swivels around the driver's seat, keeping all the colour-coded controls in easy reach.

It's a far cry from the days of the wartime Fordson driver who only had to ensure that the pointer on the calormeter stayed in the green.

The working environment for the tractor driver in the 1990s — the 260hp John Deere 8400 fitted with the CommandView cab.

Inside the John Deere CommandView cab, the operator has all the controls and monitors within easy reach. The CommandArm armrest control module is on the right of the seat.

1.   *The first British tractor to be produced in any significant numbers was the Ivel, conceived by pioneer inventor, Dan Albone. The original Ivel tractor, built up from his 1902 prototype, is shown driving the works in Biggleswade, Bedfordshire, while the factory boiler is cleaned for maintenance.*

2.   *Ivel Agricultural Motors Ltd was formed in 1903 and regular demonstrations of the tractor were held throughout Bedfordshire. This Ivel tractor is shown at one such demonstration in 1906 when, pulling two binders, it managed to cut seventeen acres of corn in six hours. The thirty-gallon cooling tanks on the later production tractors were fitted with longer vapour pipes to let out the steam and minimise the risk of the driver being splashed with hot water.*

3.   The two-cylinder horizontally-opposed engine powering this 1904 Ivel tractor gave out 24hp at the belt pulley — hardly sufficient to drive this Clayton & Shuttleworth threshing drum. Rated at 13hp at the drawbar, it weighed 37cwt and had one forward and one reverse gear.

Previous page

4. The celebrated engineering company, Marshall's of Gainsborough, entered the tractor field in 1905 with a prototype machine designed for them by Herbert Bamber, chief engineer for Vauxhall Motors at Luton. Powered by a twin-cylinder engine which started on petrol then ran on paraffin, the prototype tractor is seen driving a Marshall threshing drum in 1907.

5. Following extensive trials with the prototype tractor, Marshall's introduced a production tractor, known as the Class 'A', at the 1908 Smithfield Show. This example, working in southern Ireland in 1911, developed some 30hp and weighed nearly four-and-a-half tons. Marshall's built a number of larger machines for export before shelving production during the First World War.

6.   This 'agricultural petrol motor-tractor' was introduced in 1904 by the well-known plough and implement manufacturers, Ransomes, Sims & Jefferies of Ipswich, and powered by a four-cylinder 20hp engine driving through a three-speed gearbox. Ransomes claimed the tractor would plough five acres in ten hours with the three-furrow plough shown. The company received few enquiries for the machine and any plans to put it into production were soon dropped.

7.   H P Saunderson of Bedford, like most pioneer British tractor manufacturers, built a number of larger machines aimed at the export market. This 50hp model, seen working with a Ransomes plough at the Baldock Tractor Trials in August 1910, was designed for use in the colonies.

8.   The Eagle tractor, built in Appleton, Wisconsin, from 1916 to 1922, was one of many American machines imported into Britain during the First World War to help increase food production. The Eagle was supplied by the British American Import Company of Haymarket, London.

Facing page
9.   An Eagle 12-22 tractor with an early La Crosse chain-lift plough in the capable hands of two First World War Land Army girls.

Previous page

10.   Foster-Daimler road tractors, used to haul supplies of heavy field gun parts from the French coast to Paris during the First World War, are shown lined up in 1915. Powered by 105hp Daimler petrol engines, ninety-seven of the machines were built for the British War Office by William Foster & Co. of Lincoln, who later that year developed the first tank, initially using the same engine.

11.   Ivel Agricultural Motors never really recovered from the premature death of its founder, Dan Albone, in 1906. Having failed to implement any radical new designs of its own, the company imported the American Hart Parr 'Little Devil' 25-22 tractor, and sold it as the Ivel-Hart. This machine, seen ploughing in December 1916, had a two-cylinder, two-stroke paraffin engine powering a unique single drive-wheel.

12.   Built from 1916 to 1924, the American Big Bull tractor was marketed in Britain by the Harvesting &
Implement Department of the London store, Whiting (1915) Ltd of Euston Road. Sold as the Whiting-Bull, the tractor
is shown working with a three-furrow Ransomes riding plough. The company claimed the machine 'Could be
mastered in a week by any farm hand — male or female.'

13.   A British demonstration of the American Standard Detroit 10-20 tractor working with a Ransomes plough. Built in Detroit, Michigan, in about 1916, it was powered by a four-cylinder petrol engine and driven by two large drive-wheels, each 27in. wide.

14. The ingenuity of Britain's pioneering tractor manufacturers was matched by an enterprising Lincolnshire farmer, Henry Burtt, who built himself this impressive machine in 1916. Burtt's tractor was assembled in Foley's Agricultural Engineering Works at Bourne, using redundant steam engine parts and a 50hp, six-cylinder Napier car engine, making it a powerful machine for the time.

15. The Crawley Agrimotor was built at Saffron Walden in Essex by two brothers from 1913 to 1924. Fitted with a four-cylinder petrol-paraffin engine of 30hp, it sold for around £500. This example was photographed working near Royston, Hertfordshire, in 1924.

16, 17.   The American Parrett 12-25, built in Chicago, was sold in Britain as the Clydesdale tractor by R Martens & Co. Ltd of London. Powered by a cross-mounted, four-cylinder petrol engine, the tractor was shown at the 1917 Highland Tractor Trials in Scotland where it was reported to be 'an effective machine of strong and reliable construction' and 'capable of doing good work'. Parrett tractors were built and marketed in Canada under the Massey Harris name from 1918 until 1923, when production ceased.

18.   International Harvester's Mogul 8-16, produced from 1914 to 1917, was the first tractor from this famous Chicago company to be imported into Britain. The larger Mogul 10-20 model, introduced in 1917, is seen on trial with an Oliver plough. The tractor is fitted with the company's 'special steering device'.

19.   An International 8-16 Junior working with a binder in Lincolnshire during the 1920s. Priced at £300 in 1919 and aimed at the smaller farmer, this compact tractor, weighing only 36cwt, became very popular on the fenlands of eastern England. The radiator mounted behind the engine would keep the driver warm in winter and hot in summer!

20.   Renowned for its simple construction, strength and reliability, the International Titan became one of the most successful early American tractors, prior to the introduction of the Fordson. Over 78,000 were built between 1916 and 1922, and some 3,500 machines found their way on to British farms. This winter threshing scene shows a Titan 10-20 driving a Clayton & Shuttleworth drum near Peterborough.

21.   A petrol Clayton tractor pulling a Ransomes RYLT riding plough at the 1919 Lincoln Tractor Trials held at South Carlton. This early British track-layer was built with limited success from 1916 until 1928 by Clayton & Shuttleworth Ltd, the well-known firm of steam engine builders who were at one time the largest employers in Lincoln. The agricultural side of the company was acquired by Marshall's in 1930.

22.   The Maskell Motor
Cultivator was also shown at
the 1919 Tractor Trials.
Invented by J W Maskell of
Tillingham in Essex, it was
powered by a four-cylinder
25hp engine. Designed for light
market-garden work, it was
priced at £400 complete with
the plough.

23.   Advertised as 'The New
Saunderson Patent Super
Light-Weight Tractor', this
machine, fitted with a V-twin
petrol-paraffin engine, was a
last-ditch attempt by the
Bedford company to revive
falling tractor sales. Rated at
12/20hp, it was priced at £195
when it was introduced in
1922. The tractor is shown
working with a Saunderson
plough.

24.   *Many early tractors were employed in road haulage. This 30hp tractor, built at the Lincolnshire Ironworks premises of Martin's Cultivator Company of Stamford from 1919 to 1923, is shown in use with a local removal contractor.*

25.   The Peterbro' tractor first appeared at the Royal Agricultural Society's Aisthorpe Tractor Trials of 1920. Manufactured for over ten years by Peter Brotherhood Ltd of Peterborough, it was powered by a Ricardo patent petrol-paraffin engine developing 30-35hp. Most were exported to Australia and New Zealand. The tractor is shown hauling a heavy load to Peterborough station on a Great Eastern Railway wagon.

26. The first Model 'F' Fordson left the Brady Street and Michigan Avenue plant in Dearborn on 8 October 1917, and during part of the 1920s it accounted for 75 per cent of tractor production in the USA. By the time production ceased, in 1928, nearly 740,000 Model 'F' tractors had been built in the USA and Cork, Ireland, very many of which were sold to British farmers, including this 1918 example seen near Spalding in Lincolnshire.

27. A Model 'F' helps with the groundwork on a building site.

FORDSON TRACTOR
NG A PORTION OF THE PLANT AT THE WORKS OF
ESSRS JOSIAH PARKES & SONS LTD.
WILLENHALL
DURING COAL STRIKE 1926
UPPLIED BY REGINALD TILDESLEY.

21234

28.   The introduction of the Fordson was seen as an opportunity to bring cheap machine power to the masses, not just to the farmer, but also in factories and industry. A Model 'F' provides stand-by power for a Staffordshire works during the 1926 coal strike.

29. A Clayton tractor, modified and waterproofed for the Royal National Lifeboat Institution by F A Standen & Sons Ltd, undergoes water trials in the Great Ouse at St Ives, Cambridgeshire.

Below & facing page
30, 31. The Clayton was the first tractor to be used by the RNLI to launch lifeboats — a task previously done by horses. Initial trials were carried out at Hunstanton, and one of the first machines went into service at Skegness. Apparently, the tractors became difficult to steer, as sea water played havoc with the steering brakes.

32. The Irish Model 'N' Fordson was built at Cork from 1929 to 1932. Note the toolboxes in the rear mudguards on this early model which were tapered on the production tractor.

33. Roadless Traction of Hounslow built many tracked conversions of steam and motor vehicles during the 1920s and 1930s. The 'Super-Sentinel' Roadless tractor, introduced in 1924, was an attempt to harness steam for direct traction. This prototype was sold to Kenya.

Facing page
34. One of many Morris Commercial trucks that were equipped with Roadless half-tracks waits to be loaded with sugar beet.

35.   An American Wallis 20-30 tractor hauls a 12ft cut
Massey Harris 9C combine harvester on trial in
Hampshire in August 1929. The Canadian Massey
Harris company bought the rights to the Wallis tractor
in 1928.

Facing page
36.   A 48in. tread orchard model of the Massey Harris
General Purpose tractor cultivating a hop-yard in Kent.
Built from 1936 to 1938, this early four-wheel drive
tractor was not a commercial success. Probably too
advanced for its day, it was under-powered with its
22hp Hercules engine, difficult to steer and required two
sets of road bands to move it from field to field.

Facing page, top 37.   Few manufacturers could match Caterpillar's dominance of the crawler market, and by the 1930s the American company's products were gaining universal acceptance in Britain. This rare photograph, taken at the 1930 World Tractor Trials held at Wallingford in Oxfordshire, shows (from left to right) Caterpillar 'Sixty', 'Thirty', 'Twenty', 'Fifteen' and 'Ten' models, all working with Ransomes ploughs.

Facing page, bottom 38.   A pair of Caterpillar 'Twenty' tractors working with Ransomes ploughs and furrow presses in Cambridgeshire. Built from 1927 to 1931, the 'Twenty' was the first new design brought out by the Caterpillar Tractor Company, which was formed in 1925 following the merger of the firms of Benjamin Holt and Daniel Best. Only a handful were sold in the UK.

Above 39.   Seen discing some reclaimed fenland, the 15hp 'Ten' was the smallest model Caterpillar built, and was in production from 1928 to 1933.

SWANSCOMBE URBAN DISTRICT COUNCIL

40. Harvesting a badly laid crop of wheat in the Fens with a binder drawn by a 1931 International 10-20 tractor. Although more expensive than the Fordson, the 10-20, recognised for its sturdy construction, reliability and dependable power, became universally accepted by Britain's farmers.

41. An industrial version of the International 10-20 was available for general haulage and road transport work. This example, equipped with an Eagle trailer, was used for municipal work in Kent.

42. A South Lincolnshire ploughman with his 1934 International 22-36 tractor. Introduced in 1929, the 22-36 was an uprated version of the 15-30. It remained in production until the W30, WK40 and WD40 tractors came out in 1936.

43.   Marshall's returned to tractor production and introduced its line of single-cylinder diesel machines with the 15/30, first seen at the World Tractor Trials at Wallingford in 1930. A production model is shown winter ploughing in the 1930s.

44.　A 1935 Caterpillar 'Twenty-Two' ploughing in Lincolnshire. This was probably the most successful petrol-paraffin model built by the company, and very many were sold in the UK.

45.　Ploughing in straw is not a new idea — demonstrated by this Caterpillar 'Twenty-Eight' working with a Ransomes 'Jumbotrac' in the Fens during the 1930s. Built for only two years, from 1933, very few of these tractors were sold in Britain.

46.   Pedestrian-controlled machines, known as 'walking tractors', evolved to meet the needs of horticulture and the smaller farmer. This 1930 Monro-Tiller, working in cider apple orchards near Wisbech, Cambridgeshire, was powered by an engine mounted inside the single drive-wheel.

47.   The Marshall 15/30 was followed by the 18/30 in 1933 — as seen here ploughing at Petworth in Sussex in 1936. Note how the driver has streamlined himself for speed by turning his cap back to front.

48.   Two-cylinder John Deere tractors were first imported into Britain by F A Standen & Sons Ltd of St Ives, Cambridgeshire, in 1930. This model 'BR' is being demonstrated by the company near Papworth in 1937.

49.   A 1934 Fordson lifting sugar beet on the Fordson Estates in Essex. Production of the Model 'N' was transferred to Dagenham in 1932, and the tractor was painted in a distinctive dark blue and orange livery, reputedly to match the colours of the Essex farm wagons.

50.   Caterpillar's diesel-powered 'RD' series was introduced in 1936. The RD7 replaced the 'Diesel Fifty', eventually becoming the D7 in 1938. This RD7, ploughing in Lincolnshire with a three-furrow Ransomes 'Junotrac', is fitted with electric lighting and a Leverton cab.

Top
51. A Caterpillar drawing twenty-one furrows at one pass in Lincolnshire in 1938. The Ransomes ploughs consist of two Hexatracs, a Quintrac and a Multitrac.

52. A Caterpillar D7 9G working a Ransomes two-furrow 'Twinwaytrac' balance plough. Only five two-furrow and two single-furrow 'Twinwaytracs' were made, and all went to the same area in Lincolnshire.

53. While the Case 'L' tractor became a popular machine with British farmers, the American company's combine harvesters would be an uncommon sight in the UK. Following many years' experience gained producing threshing machines, J I Case of Racine, Wisconsin, introduced three models of combine, the 'A', 'P' and 'H', during the 1920s. The model 'L' tractor first appeared in Britain at the 1930 Tractor Trials, rated at 26-40hp and priced at £348.

54. A 1937 Case 'CC' hoeing cabbages in Bedfordshire with a Ransomes toolbar. The 'CC', a rowcrop version of the model 'C' tractor, incorporated the 'Motor-Lift' system — a simple worm and gear unit driven off the power-take-off shaft to raise and lower the implements.

55. *This Cleveland Cletrac '40' Diesel 46-60 crawler, photographed with a Ransomes Hexatrac plough near Bourne, Lincolnshire in 1936, would be an impressive machine at the time. It was fitted with a Hercules six-cylinder engine, Leece-Neville electric starting and lighting equipment and weighed just over five tons.*

56, 57. The first Roadless conversion of a Fordson followed a request by Margate Corporation for a tractor capable of travelling on loose sand to haul seaweed off the beach, a task previously undertaken by horses. The Fordson-Roadless tractor, fitted with rubber-jointed tracks, is shown on demonstration in April 1930, easily hauling a three-ton load of wet seaweed from Margate sands to the town, even when the wagon wheels sank to a depth of nine inches in places.

Facing page
58. This Bristol tractor, seen hauling a London trolley-bus into position for the 1937 Commercial Vehicle Exhibition, was one of four Roadless-tracked machines used at Earls Court for moving exhibits. The Bristol was fitted with a front castor wheel to aid stability over uneven surfaces.

59. During the 1940s, some of the first forage harvesters were imported from the USA for chopping and loading silage, including this Fox-Rivers machine being driven by an Oliver 80 Standard tractor in Yorkshire.

60. A Minneapolis Moline ZTN planting cabbages near Hitchin, Hertfordshire, in 1948. Built and imported from the USA in 1943, the tractor was allocated to the Bedfordshire War Agricultural Committee and was originally supplied on steel wheels.

Facing page
61. The Case LA, introduced in 1941, was a superb tractor — well-built and well-liked. The tractor in the photograph has been fitted with Roadless 'DG' half-tracks and is ploughing with a five-furrow Ransomes Hexatrac. The special extension bars on the plough are probably to help bury straw or trash. In the background, a Fordson working with an early Catchpole beet harvester is happily boiling away.

62. One of many Caterpillar D2 tractors imported during and after the Second World War — a 1948 'U' model planting potatoes in Lincolnshire.

63. Two wide-gauge Caterpillar D2 5J models ploughing in the Fens. The 'J' model was manufactured from 1938 to 1947.

64.   With the main dealers, Levertons, based in Spalding, it was not surprising that so many Caterpillars went to work in Lincolnshire and the surrounding fenland. This early D4, built around 1939, helps out with a farm fire by dragging a burning stack away from the yard.

65. A 1946 Allis Chalmers HD10 crawler ploughing with a Ransomes Marquis five-furrow plough near Dowsby in Lincolnshire in 1949. The HD10 was powered by a General Motors 4-71 four-cylinder, two-stroke diesel engine developing 86hp.

Facing page

66. Harvesting during the Second World War with an International Harvester 31T combine drawn by an International TD6 crawler — in stark contrast to the old methods demonstrated in the background as sheaves of corn cut by the binder are brought in by horse-drawn wagons and stacked ready for threshing.

Many weird and wonderful machines were produced during the 1940s as the drive to mechanise agriculture allowed free rein to the inventors' imagination. Some of these early attempts became the precursors of the machines we know today, while others were sent back to the drawing board, or simply forgotten.

67.   An early McConnel hedge-trimmer fitted to a Fordson tractor working in Cambridgeshire. Power for the blade was provided by an auxiliary Petter engine mounted on the balance bar.

68.   A self-propelled 'unit-binder'. Designed by a Mr Hutchinson, it was built by the National Institute of Agricultural Engineering using parts from a Canadian-built McCormick International binder fitted to an Allis Chalmers WC tractor.

69.   *The forerunner of the front-end loader? The Painter muck-loader mounted on a Model 'N' Fordson; the loader was raised by a winch arrangement driven off the tractor's belt pulley. The big drawback to the system was that with all the weight so far forward, the tractor had no grip and it was impossible to work in slippery conditions.*

The John Deere at work.

70.   A 1943 John Deere 'BN' fitted with a toolbar for ridging potatoes, work for which its three-wheel rowcrop configuration was ideally suited.

71.   Another Model 'BN' draws out the rows and spreads No 1 National Compound fertiliser for potatoes.

Facing page
72.   Two John Deere 'BN' tractors, built around 1941, struggle to deliver a trailerload of sugar beet to the Spalding factory in the January snow.

73.   David Brown of Meltham, Huddersfield, launched its first tractor, the
VAK 1, in 1939. During the Second World War, the company produced a
number of industrial aircraft-towing tractors for the Air Ministry, known as
the VIG 1. After the war several of these were fitted with a front-mounted
pulley and sold as threshing tractors for peacetime use — as shown in this
rare photograph of a 1945 model driving a Marshall drum in Yorkshire.

The David Brown Cropmaster
was one of the company's best-
known tractors and nearly
60,000 were produced between
1947 and 1953. Although it
was initially available only
with a petrol-TVO engine, a
diesel version was introduced
in 1949.

74.   A David Brown
Cropmaster drilling sugar beet
with a Wild Sunit Gang Seeder.
Although this drill offered some
degree of seed metering, the
crop would still have to be
gapped and singled by hand.

75.   A Scandinavian Mern 'PL'
sugar beet harvester fitted to a
Cropmaster. The topping unit
was mounted on the front of the
tractor, and the lifter-cleaner
was drawn behind, leaving the
beet in heaps across the field to
be loaded by hand.

76.   *Improved mechanised harvesting methods came to the fore during the 1940s. A 1940 John Deere 'BN' with a four-wheel trailer collects eighteen-stone sacks of wheat from a self-propelled Massey Harris 726 bagger-model combine working in Lincolnshire. The 726 was built from 1949 in the new Massey Harris factory at Kilmarnock in Scotland.*

77.   An American-built International Farmall 'M' pulls an early Catchpole sugar beet harvester. Catchpole Engineering of Bury St Edmunds in Suffolk went on to manufacture a long line of sugar beet harvesters before being taken over by Ransomes in 1968.

78.   Harvesting potatoes in the Fens with a 1944 John Deere Model 'AN' tractor and a Johnson harvester, built at March, Cambridgeshire. The potatoes are collected by a Muir-Hill 'Powacart' — one of a number of war surplus dumpers converted by Muir-Hill for agricultural use.

The legacy of the Second World War left farmers in post-war Britain facing a call to increase productivity with only limited manpower available to them — leading not only to improvements in mechanisation, but also to the more extensive use of chemicals in arable agriculture. The practice of spraying cereals with sulphuric acid as a method of eradicating weeds became widespread from 1911, with the disadvantage that it was highly corrosive both to the sprayer and the operator. During the war, trials were carried out with di-nitro-ortho-cresol, or DNOC for short. Originally used as a slimming salt until found to be highly toxic, DNOC was discovered to be a very effective alternative

79.   On this type of sprayer the chemical was delivered by compressed air from a spray tank pressurised by a belt-driven compressor.

to sulphuric acid for selective weed control in cereals. The disadvantage with the early sprayers was that they were unable to suitably atomise the spray, and water rates of up to a hundred gallons to the acre were often required. As can be seen from the photographs, little or no protection was offered to the operators.

80. A 1940 John Deere 'BN' crop-dusting potatoes. Applying chemical powder with a duster was an often preferred alternative to the sprayer, especially for insecticides. The potatoes are probably being dusted with copper-lime powder for blight or lead arsenate for Colorado beetle.

81. A Fordson E27N Major spraying wheat in Cambridgeshire with a conventional pump-driven sprayer in 1947.

During the 1940s the grassland farmer began to recognise the benefit of increased mechanisation.

82. A 1944 Allis Chalmers Model 'U' working in Yorkshire with a Tullos-Goodall Grass Conserver. This machine had a flail-rotor which was designed to lift and lightly pulverise newly mown grass to speed up the drying process for hay.

83. A rear-mounted buckrake on a Fordson E27N Major is used to gather up grass and then transport it to the silage clamp.

1. A Lincolnshire farmer gets a demonstration of the new International B275 in 1959. Note the old Series 1 Landrover in the background. The 35hp B275 replaced the B250 and was made up until 1961.

2. A Massey Ferguson 175, fitted with Bettinson steel wheel extensions, gets to grips with some heavy clay soil in Rutland during a wet autumn in 1968. The 66hp 175 was one of Massey Ferguson's 'Red Giants' range introduced in 1964.

3. A County 1454 stirs the dust with a set of gang rolls in 1973. By that time, the company had manufactured 30,000 tractors, and was exporting to 143 countries worldwide.

4. A County 7600-Four wallows in the mud. Built in about 1976 and based on the 97hp Ford 7600, this unequal-size four-wheel drive tractor is seen fitted with a Brown's loader.

Facing page
5. Introduced into the UK in 1975, the Massey Ferguson 1505 articulated tractor was powered by a V8 Caterpillar 3208 engine delivering 180hp. The tractor in the photograph has been fitted with an M & W turbocharger and boosted to over 200hp.

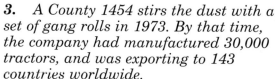

6. *Discing burnt stubble with a Ford FW-30 powered by a 295hp Cummins V8 engine. The giant FW-30 and FW-60 pivot-steer tractors were built for Ford by the American Steiger company of North Dakota, and introduced into the UK market in 1978. Approximately 145 were sold in Britain over the next five years.*

7. *A massive John Deere 8850 on dual wheels makes an impressive sight ploughing up stubble. Powered by John Deere's own 370hp, turbocharged and intercooled V8 engine driving through a sixteen-speed Quad-Range transmission, the 8850 was the largest of three articulated four-wheel drive tractors introduced by the company in 1981.*

Left
8. *Twenty years apart — a 1960s John Deere 5020 works alongside a 1980s John Deere 4440 near Kings Lynn in Norfolk.*

Below
9. *A 135hp Fiat 1380DT working with a set of Opico Bush Hog disc harrows in 1981. The 1380 was introduced in 1979 — a year in which Fiat Trattori's production range comprised more than fifty different models of tractor.*

Facing page
10. *A 1980 165hp Caterpillar D6D Special Application tractor, fitted with a quiet cab and three-point linkage, ploughing with an eight-furrow Dowdeswell.*

11. A 1989 Case International 1056XL ploughing with an Opico Square Plow. The 1056XL was powered by a six-cylinder engine developing 105hp, and was fitted with a 16x8 synchromesh transmission and the company's 'Sens-o-draulic' hydraulic system.

12. Unveiled in Geneva in November 1994, Caterpillar's new, smaller Challenger tractors are designed to be used for drilling and spraying operations as well as ploughing and heavy cultivations. Powered by Caterpillar's own 3116 air-to-air-after-cooled engines, the machines share several parts with the New Holland Ford 70 Series tractor. The 205hp Challenger 35 is seen on its first UK demonstration in Lincolnshire in December 1994, ploughing with a Dowdeswell DP160 seven-furrow plough.

13. The top model in JCB's Fastrac range of high speed tractors was launched at the 1994 Royal Show. Powered by a 170hp six-cylinder turbocharged and intercooled Cummins 6BTA engine, the Fastrac 185 is seen with a Dowdeswell plough.

Facing page

14. New Holland's Fiat G240 tractor, powered by a six-cylinder turbocharged 240hp engine with air-air-intercooling driving through an 18x9 powershift transmission. The G Series and the identical Ford 70 Series tractors were released in February 1994.

15. Massey Ferguson launched two new tractor ranges at the Paris SIMA Show in February 1995 — the 80hp to 120hp 6100 Series, and the larger 8100 Series of 135hp to 200hp machines. The 155hp 8130 is powered by a six-cylinder turbocharged Perkins engine and equipped with a 32x32 Dynashift transmission.

16. The 130hp John Deere 6900 was added to the company's 6000 Series of tractors in October 1994, powered by a six-cylinder engine and fitted with sixteen-forward/twelve-reverse PowerQuad transmission giving a 30kph top speed.

17. *Massey Ferguson announced another new tractor in early 1995 — the 240hp 9240, fitted with a six-cylinder turbocharged and intercooled Cummins engine, and an 18x9 powershift transmission. Built in collaboration with MF's new parent company, AGCO, it is based on the American White tractor range. The tractor's rear axle is made in Yorkshire by the David Brown engineering group.*

84.   An International TD6 crawler with a set of gang rolls is used to consolidate the silage clamp. The American TD6 was the smallest of a range of three petrol-start diesel crawlers introduced by International Harvester in 1941.

Even though the combine was becoming a familiar sight in Britain, it was an expensive machine, and many farms, especially the smaller ones, preferred to keep to traditional harvesting methods.

85.   *Mother can come too — if you have the advantage of a David Brown with a double seat. A 1949 Cropmaster draws the binder during the wet harvest of 1954.*

Facing page
86.   *Threshing in Lincolnshire on a winter morning with a Foster drum driven by a 1949 Field Marshall. A Fordson Major, fitted with a Hesford winch, was needed to help position the drum as the damp ground conditions had proved too much for the Marshall.*

Facing page

87.   Harvesting in Norfolk with a Grain Marshall 568 combine towed by a 1946 Fordson E27N Major tractor.

Above

88.   A prototype 'Albion' combine on trial with a 1949 Fordson Major. The combine was built by Harrison, McGregor and Guest Ltd of Leigh in Lancashire, an old-established firm of implement manufacturers who were eventually taken over by David Brown in 1955.

Right

89.   Baling in Lincolnshire with a 1945 Fordson E27N Major and a McCormick International B45 baler.

Introduced in 1945, the E27N Major was the long-awaited replacement for the ageing Model 'N' Fordson. The Fordson tractor had always been a popular choice for conversion by the specialist industrial machinery manufacturers, and the E27N was no exception.

90.   A light load for a 1948 E27N Major, fitted with a Leeford mule-dozer and Rotaped tracks, working on a Middlesex building site.

91.   Shunting railway wagons with an E27N-based 'Buffer' tractor built by Chaseside Engineering of Hertford.

Facing page
92.   In 1946, the Labour government passed the New Towns Act, authorising further development and outlining plans for the creation of up to twenty new towns. Construction machinery manufacturers were faced with a post-war boom in sales for their products, and the E27N Major proved a reliable and economical source of power for their products. Two Chaseside 'Hi-Lift' shovels are seen at work on a building site at Orpington in Kent.

93.   A 1948 Ferguson tractor with a Ferguson three-ton hydraulic tipping trailer. When Harry Ferguson brought out the Ferguson TE20 tractor in 1946, he introduced the farmer to a whole new system of mechanised agriculture with his matched range of implements and equipment.

Facing page
94.   This Priestman Cub 'V' dragline excavator, mounted on its low-loader trailer in Hull, would be a heavy load for the 1954 Ferguson TEF20 diesel industrial tractor.

95.   *An early Fowler Mark VF crawler, dating from around 1947-1949, with a Martin's of Stamford cultivator. Essentially a Marshall Series II tractor on Fowler track running gear, it was the result of a 1946 merger between the two companies. Built at Leeds, it is obviously photographed in a far warmer part of the world. Like many British tractor manufacturers, Fowler-Marshall exported many of their products.*

96.   A British-built Massey Harris 744D tractor fitted with a hydraulic lift uses a Ransomes TS55 deep-digger plough to break up some moorland turf. An initial batch of 744PD tractors was assembled at Manchester in 1948, primarily for the groundnut scheme, before full production was transferred to Kilmarnock the following year. Power was provided by a six-cylinder Perkins P6 diesel engine rated at 42bhp.

97. The Gunsmith light tractor, introduced in October 1948 by Farm Facilities Ltd of Marylebone Road, London, was one of several new lightweight machines brought out during the late 1940s, aimed at the smaller farmer and market-gardener.

98. The Vivian Loyd company of Camberley, Surrey, built an agricultural crawler tractor in 1946, based on its wartime military Bren gun carrier and powered by a 30hp Ford V8 petrol engine. The prototype is seen working with a four-furrow Cockshutt plough. The machine was in full production by January 1946, priced at £495, and led to several variations, culminating with the Loyd Dragon in the 1950s.

99. A 1955 Bean Rowcrop Tractor is used to hoe a sixty-acre field of sugar beet in Oxfordshire. Powered by an 8hp side-valve Ford car engine, the machine was originally manufactured from 1945 by the Humberside Agricultural Products Company to the design of local grower, Mr Bean. The tractor was later built in the Smithfield Ironworks of Thomas Green & Son Ltd of Leeds.

Facing page

100.   A 1948 Field Marshall Series II ploughing in Somerset with a Ransomes RSLM Major TS29 plough. Marshall's of Gainsborough introduced the Field Marshall range of single-cylinder diesel tractors in March 1945. The Series II was built from 1947.

101.   A Field Marshall Series III ploughing in Yorkshire in 1952. The Series III replaced the Series II in 1949. Despite declining sales, the company persevered with the single-cylinder design until 1957.

102.	Planting potatoes on a fenland farm with four Ferguson TE20 tractors in 1952.

103.	When wet July conditions prevented him from ridging his potatoes before the foliage had met in the rows, Lincolnshire farmer Ken Hatter overcame the problem by reversing the ridging bodies on the toolbar and driving his 1953 Allis Chalmers Model 'B' tractor backwards.

104.   *Lifting potatoes in Essex with a trailed Ransomes No 41 potato spinner pulled by a Fordson E27N tractor. Ransomes had a long association with Ford, and developed a range of 'FR' implements for use with the Fordson Major tractors.*

Above and facing page
105, 106.   A 1950 Massey
Harris 744D tractor in
Hampshire with a Massey
Harris 701 baler, powered by
an air-cooled Wisconsin four-
cylinder auxiliary engine.

107.   A rare rowcrop version of
the Massey Harris 745 tractor
baling hay in Warwickshire
with a 701 baler driven by an
Armstrong Siddeley diesel
engine. The new 745 tractor
replaced the 744 in 1954, and
was powered by the four-
cylinder Perkins L4 engine
developing 44.6bhp.

108.   After the Cropmaster's long production run ended in 1953, David Brown introduced five new tractors. The 25 and 30 models had petrol-paraffin engines, while the 25D, 30D and six-cylinder 50D were diesel powered. All had the advantage of a six-speed gearbox. The 31hp 25 is shown with an Albion spring-tine cultivator.

109.   Preparing a seedbed in 1951 with a Nuffield DM4 tractor near Peterborough. Soon after the Nuffield Universal was launched in 1948, it became obvious that farmers wanted a diesel version. Unable to develop its own BMC diesel engine quickly enough to meet demand, Morris Motors used the 48hp Perkins P4 (TA) power unit for the DM4 from 1950 to 1954.

110.   When the new Fordson Major was launched in November 1951, it was the diesel version that was to have the greatest impact on British farmers. Thousands were sold, and many are still at work all over the world. The photograph shows a 1955 Diesel Major drawing a trailer alongside a Lundell Super 60 forage harvester pulled by a 1958 Fordson Power Major.

111.   Among the many conversions based on *Fordson Major* tractors was this *1955 County Full Track Mark II, seen planting potatoes in south Lincolnshire with a Robot planter. County built crawlers from 1948 until 1965. The company then concentrated on manufacturing four-wheel drive tractors.*

112.   A David Brown 900 diesel on demonstration in Lincolnshire with an Albion 'B' Series three-furrow plough in December 1956. The 900 was built from 1956 to 1958, and the diesel version was rated at 40hp.

Facing page
113.   Ploughing in Oxfordshire with a 1958 Nuffield Universal Four. The 40hp BMC diesel-powered Universal Four was made from 1957 to 1961.

114. A Ferguson TE20 drilling sugar beet with a precision seed drill which did away with the laborious task of gapping and singling the plants by hand.

115. Steerage-hoeing oilseed rape in south Lincolnshire with a 1958 Fordson Dexta. The little Dexta was Ford's answer to the Ferguson, and was produced in various forms from 1957 to 1964. Its three-cylinder diesel engine was built by Perkins to Ford's specifications using Ford castings.

116.   An International B250 ridging potatoes near Moulton Chapel in Lincolnshire. International's small 30hp tractor was built at the old Jowett car factory in Bradford from 1956 to 1958.

117. David Brown built a number of crawler tractors during the 1950s. The photograph shows a 1953 Trackmaster 50TD powered by a six-cylinder diesel engine of 50hp.

118. A Fiat 55C is used to alter a watercourse and bulldoze a new river bed. Fiat crawlers were unusual in having a steering wheel in place of the normal steering clutches. Built at Modena in Italy, they were sold in Britain from 1950 through MacKay Industrial Equipment Ltd of Feltham, Middlesex.

Facing page
119. Clearing old hedges with a Bristol 22 crawler fitted with a Bray angle-dozer and a Hesford winch. Built from 1952, the Bristol 22 was powered by a three-cylinder Perkins diesel engine.

From the late 1940s onwards, many manufacturers rose to meet the challenge of mechanising horticulture.

120.   The Colwood Rotary Hoe cultivating between daffodil rows near Spalding, Lincolnshire. Introduced by Dashwood Engineering of London in 1949, it was powered by a Villiers engine.

121.   The specialist rowcrop David Brown 2D tractor was built from 1956 to 1961. Aimed at the market-gardener, it had a two-cylinder diesel engine and the lift was operated by compressed air.

122.   A Clifford 'A' Mark III rotary cultivator working in the snow near Wisbech, Cambridgeshire. Built from 1949 by Clifford Aero & Auto Ltd of Birmingham, it was driven by a single-cylinder, four-stroke petrol engine developing 6hp.

123.   This mini-tractor was built and designed in 1962 by W J Clark of Upwell, Cambridgeshire, for working in apple orchards. It was powered by a 10hp Petter PC2 diesel engine.

124.   Many farmers still favoured the old harvesting methods well into the 1960s. Threshing in Lincolnshire in August 1966 with a Clayton drum driven by a 1953 Fordson Diesel Major. Another Fordson brings in the sheaves.

Facing page
125.   Harvesting oats in Perthshire in 1960 with two Massey Ferguson 35 tractors and binders.

The GBW beet harvester, made at Maxey, near Peterborough, was typical of the lightweight single-row machine favoured by farmers during the 1960s, as mechanised sugar beet harvesting took over.

126. *A GBW Beet King, drawn by a 1950 David Brown Cropmaster, delivers beet to a scotch cart pulled by a Massey Ferguson 35.*

127. *Another David Brown, this time a 1953 model 30, hauls a three-ton trailer alongside the smaller GBW Atom 25 beet harvester behind a 1957 International B250.*

128. A 1955 Nuffield Universal catches the sugar beet lifted by the GBW Atom beet harvester. The four-wheel trailer, which did not transfer any weight on to the drawbar, was difficult to move in sticky conditions, and was normally only used for road haulage to the factory.

129.   A Massey Ferguson 65 is fitted with cage-wheels to help it cope with harvesting peas under wet conditions with an early Mather and Platt mobile pea viner. These machines evolved as the pea freezing industry became more established during the 1960s. Inside the viner, the windrowed swath was fed to a drum where the peas were separated from the haulm. After screening, the cleaned peas are collected in a hopper which would hydraulically tip into a trailer or lorry ready for delivery to the factory.

130.   A July harvest. A 1963 Fordson Super Major collects the corn from a crop of winter barley harvested by a 1964 *Clayson M103 combine. Imported by Bamfords of Uttoxeter, Clayson combines were built in Belgium by Leon Claeys, manufacturers of the first European self-propelled combine in 1952. Claeys were taken over by New Holland in 1964.*

131.   An International B450 on site-clearance with a Leeford mule-dozer. Fitted with the 55hp BD264 engine, the B450 enjoyed a long production run from 1958 to 1970. It was built at International Harvester's new British factory that had opened in 1949 on a sixty-acre site at Wheatley Park, Doncaster.

132.   Crawler tractors were also built at the Doncaster factory. The International BTD6 was very similar to the American TD6 tractor, but again had the British BD264 engine. The plough is a Norfolk Bonnell, specially developed for the BTD6 by Continental Farm Equipment Ltd of Salhouse, near Norwich, using imported French parts.

133.   An old coppice hedgerow in Lincolnshire is pulled out by an early model International BTD640. Note the old Agricastrol can which would hold a supply of diesel ready to start the bonfire. International crawlers were manufactured in Britain up to the mid-1970s.

134.   A Track Marshall 70 working with a Doe heavy-duty seven-tine cultivator attached to a Doe tool carrier. Built by Ernest Doe & Sons Ltd of Ulting, Essex, the tool carrier allowed crawlers to use heavy-duty three-point linkage implements.

135.   A Fowler Challenger 33 uses a large mole-plough to drag out bog oaks in Lincolnshire. The 33, introduced in 1958 and fitted with a 125hp Leyland 680 engine, was the culmination of the Challenger line and the last tractor to bear the Fowler name.

Even in the 1960s, hand labour and horse power still had an important place on the farm.

136.   October 1965, and tulip bulbs are scattered down the rows in one of the many bulb fields around Spalding in Lincolnshire. The tractor on the cart is a 1960 David Brown 950 Implematic.

137.   December 1962, and horse power and tractor stand together as sugar beet is unloaded by hand from two scotch carts or tumbrils.

138.   October 1965, and horses wait patiently for their carts to be filled with potatoes hand-picked behind a 1963 *International 414 tractor with a Johnson hoover. The horse remained in service on many south Lincolnshire farms long after it had disappeared from other parts of the country.*

Left and facing page

139, 140. A big job for little tractors. A Massey Ferguson 65 and a Fordson Diesel Major come to the rescue of a stranded oil tanker on the River Nene, near Sutton Bridge, Lincolnshire, in April 1966. The coaster had run aground after hitting the river bank and had become lodged across the river. The Massey had been fitted with cage-wheels to increase its traction, but even with over 100hp between them, it was perhaps a little optimistic to expect the tractors to move the tanker until the tide came back in. Note the period weather cabs on the tractors — the Massey has the benefit of a Winsam, while the Fordson is fitted with a Sun-Trac cab.

141.   An industrial use for an agricultural tractor — an Allis Chalmers ED-40 fitted with a Leeford mule-dozer. The ED-40, built from 1960 to 1969, was the last tractor to be made at the Allis Chalmers works at Essendale, near Stamford. It was fitted with a Standard 23C diesel engine that was a notoriously bad starter and probably instrumental in the tractor's disappointing sales record.

142.   An agricultural use for an industrial tractor — a Wiltshire fencing contractor uses a JCB 2 fitted with a Danuser hydraulic post-hole borer. JCB machines were originally based on Fordson and, later, Nuffield and Leyland skid-units.

143. *Berkshire farmer, Pat Saunders, gained more power and extra traction by linking two tractors together — in this case, a Fordson Major and a Nuffield Universal — with all the controls in easy reach of the driver of the leading tractor. His system was eventually marketed by Paramount Engineering of Coventry as the 'Dual Tractor Kit'.*

144. *Another way to increase traction was to fit your John Deere 4020 with Arps flexible half-tracks, marketed in Britain by Opico of Bourne, Lincolnshire. The 4020 was rated at 106hp and nearly 200,000 were built.*

Above and facing page
145, 146.   *Ploughman Willy Chatterton remembers coping with some dreadful conditions when this fifteen-acre pasture on Cowbit Wash, Lincolnshire, was ploughed up for the first time in April 1963 — a wet year. The 1959 Massey Ferguson 65 was fitted with a single cage-wheel for extra traction, and was assisted on a chain by a 1952 Fordson Diesel Major. The two tractors struggled as water fountains sprang up in the field, and more than once had to be pulled out by a Caterpillar D7 after getting into difficulties. The field was planted with potatoes that spring.*

147. A 1966 Ford 5000 is employed in silage making, picking up wilted herbage with a New Holland double-chop forage harvester. Renowned as a sturdy and reliable tractor that would take a lot of abuse, the 5000 became a popular tool for the contractor during its ten-year production run.

148. Doe 130 and 'Triple-D' tractors would cope with most conditions, but are brought to a standstill on maize stubble in Essex after heavy autumn rains.

149. *Following periods of heavy rain, two Doe 130 tractors, pulling Mather and Platt Super mobile pea viners and a Hesston swather, move in on part of 1,000 acres of peas grown by a Lincolnshire co-operative near Grantham in 1968. The Doe 130, based on two 65hp Ford 5000 tractors, was built in small numbers by Ernest Doe of Essex.*

150. *One of two Hesston swathers modified by the importers, Opico, to harvest Brussels sprouts for Bird's Eye in 1968. The cut stalks are elevated into a cart pulled by a Massey Ferguson 178. Hesston swathers were originally imported with Wisconsin petrol engines, but most were converted by Opico to Perkins 4-107 diesels.*

David Brown launched its Selectamatic range of 'white tractors' in 1965.

151. A 1969 David Brown 1200 fitted with Opico half-tracks in Cambridgeshire. The 1200 was the largest model in the Selectamatic range. Introduced in 1967 and uprated to 72hp in 1968, it suffered from some unreliability problems.

152. Earthing-up celery with a 1965 David Brown 990 high clearance tractor. Most David Brown tractors could be simply converted to high clearance by rotating the rear axle final-drive units and fitting an extended front axle. The conversion became very popular with the green vegetable and salad growers in the eastern counties.

153.   When the going got tough … a Yorkshire farmer coupled two David Brown 990 tractors together in tandem to cope with winter ploughing in January 1967. The combined power of the two 55hp engines gave him over 100hp to play with.

154.   Two industrial Ford 5000 tractors, fitted with Wickham Poole semi-trailers, are used to haul cement from the works. The tractor was never just confined to agriculture, and the Ford 5000 quickly established a reputation as an industrial work-horse.

155.   The Roadless Ploughmaster 65 was one of several four-wheel drive conversions based on the Ford 5000. The plough is a six-furrow semi-mounted Doe with a steerable headstock.

156.   A 1968 Muir-Hill 101 lifting potatoes in Lincolnshire with a Krakei harvester. Introduced in November 1966, the 101 was designed as a four-wheel drive machine with both agricultural and industrial uses that would also appeal to overseas timber and sugar cane growing areas. Many transmission and hydraulic parts were derived from the Ford 5000 and power was provided by a six-cylinder 108hp Ford 2704E engine.

157.   *County of Fleet, Hampshire, were the most prolific manufacturers of Ford-based four-wheel drive conversions. The 113hp 1124 tractor, seen engaged in pipeline construction with a Wickham Poole trailer, was introduced at the Royal Show in 1967, and again powered by a Ford 2704E engine.*

158. The German Fendt was one of many European makes to be imported into the UK from the 1970s. The Fendt Toolcarrier, powered by a 45hp three-cylinder air-cooled engine, was designed as a multi-purpose machine that could be used with a combination of attachments.

159. The 83hp Fendt 108S Turbomatic, seen spring cultivating in April 1975, had the advantage of an hydraulic turbo-clutch which allowed for smooth starting.

Facing page
160. The German-built International 523 tractor had a three-cylinder 52hp engine and the option of a semi-automatic Agriomatic transmission which had twelve forward and four reverse gears.

161. The Massey Ferguson 1080 was added to the MF product line in 1970. Aimed at both the European and American markets, it was powered by a 90hp Perkins A4.318 four-cylinder diesel engine developed specifically to meet the requirements of the tractor. Perkins had been bought by Massey Ferguson in 1959.

162. The North American Massey Ferguson 1135 and 1155 tractors were introduced to the UK in 1975. The 1135 was fitted with a six-cylinder turbocharged Perkins AT6.354 engine developing 135hp. Neither tractor achieved very high sales in Britain.

163.   *The massive 155hp Massey Ferguson 1155 tractor was powered by a V8 Perkins AV8.540 engine. It was probably one of the most powerful two-wheel drive tractors ever sold in the UK.*

164.   *The Leyland 482 Synchro, shown with a four-furrow Lemken plough, was fitted with the turbocharged 4/98TT engine developing 82hp and a Carraro four-wheel drive front axle. The Leyland Synchro range was introduced in 1978 and produced at Bathgate in Scotland.*

165.　A 1983 David Brown 1390 with a Hardi sprayer. Launched in 1979, the 90 Series was the last range of tractors to bear the David Brown name. The subsequent 94 Series was sold under the Case banner following a rationalisation in marketing by the US industrial conglomerate, Tenneco, owners of David Brown Tractors and J I Case. Following Case's acquisition of International Harvester in 1985, the David Brown plant at Meltham was eventually closed.

166. A Fiat 90C crawler with a Dowdeswell reversible plough and furrow-press. The 90C was fitted with a four-cylinder 98hp engine, and was built at Fiat's Cento factory from 1976.

167. The Track Marshall 155 was introduced in 1987 by the newly formed company, Track Marshall's of Gainsborough Ltd. Powered by a six-cylinder, turbocharged Perkins 6.354 engine, it is seen with a six-furrow reversible plough built by Marshall dealers, Robert Crawford & Son of Frithville near Boston, Lincolnshire.

168.   *Tractors at play — a 1982 62hp Universal U640 tractor is used to launch boats on Coniston Water in the Lake District. Universal tractors have been built since 1947 by UTB, the Romanian state tractor factory. From 1967, the factory has produced Fiat tractors under licence.*

169.   The Ford TW range of tractors was brought out in 1979, and then further uprated in April 1983 when the TW-15, TW-25 and TW-35 models were released. The TW-35 Force II, fitted with a turbocharged and intercooled six-cylinder engine developing 186hp, was introduced at the 1985 Smithfield Show.

170.   New Holland Ford's new Series 40 Powerstar tractors were announced in November 1991. Built at Basildon, the range featured new engines, new transmissions and a new cab design. The 8340 SLE, seen with a New Holland 550 forage harvester, has a six-cylinder 125hp engine and a 16x16 'Electroshift' powershift transmission.

171.   Massey Ferguson's 3065 'High Visibility' tractor was launched in 1992, with the benefit of Autotronic or Datotronic electronic control and information systems, and powered by an 85hp turbocharged Perkins engine.

172.   Czechoslovakian Zetor tractors have been sold in Britain since 1966. The four-wheel drive Zetor 7340, ploughing set-aside land in June 1995, is turbocharged to 84hp.

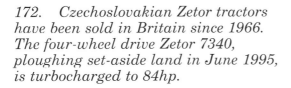

173.   A tractor for the twenty-first century? The John Deere 8300 with 230hp on tap has a price tag of over £80,000. Introduced in October 1994, the 8000 Series four-wheel drive tractors were the first in history to have their design concept patented.

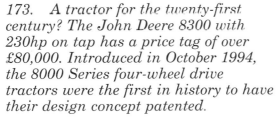

The End

INDEX

Roman numbers refer to text captions, bold to the colour section

FARMING PRESS BOOKS & VIDEOS

Below is a sample of the wide range of agricultural and veterinary books and videos we publish.
For more information or for a free illustrated catalogue of all our publications please contact:

Farming Press Books & Videos
Miller Freeman Professional Ltd
Wharfedale Road, Ipswich IP1 4LG, United Kingdom
Telephone (01473) 241122   Fax (01473) 240501

Tractors at Work Volume 1   *Stuart Gibbard*

Stuart Gibbard sets the tractor firmly in its working environment, showing in an outstanding compilation of photographs as much about the recent social history of farming as about machinery development.

Classic Farm Machinery Volumes 1 and 2

Compilations of archive video chosen and expounded by Brian Bell to show the development of agricultural machinery. Volume 1 covers 1940-70, Volume 2 1970-90. '… you will never tire of watching … I would not hesitate to recommend to anyone with an interest in farm machinery' *Farm and Horticultural Equipment Collector.*

Ford Tractor Conversions: the story of County, Doe, Chaseside, Northrop, Muir-Hill, Matbro and Bray   *Stuart Gibbard*

This magnificently illustrated book deals in depth with these pioneering British companies and their main models from the turn of the century to the present day.

World Harvesters   *Bill Huxley*

Photographs and short descriptions of a wide range of harvesters of all types from all over the world.

Fifty Years of Farm Machinery
Fifty Years of Garden Machinery   *Brian Bell*

A pair of books which illustrate and describe the course of rural machinery development in Britain from the 1940s to the present. A host of models and manufacturers are dealt with, the emphasis being on the 1950s and 60s when progress was most rapid.

Books and Videos *by Michael Williams*

Farming Press have published five books by Michael Williams and four videos in which he is the presenter. Topics include Ford & Fordson, Massey-Ferguson and John Deere as well as the bestselling **Tractors Since 1889** and the children's **Tractors: How They Work and What They Do.**

Early and More Years on the Tractor Seat
        Arthur Battelle

Two humorous autobiographical accounts of one man's life with machinery from a pre-war Fordson N and an IH W30 to the Fordson Dexta and Major.

Farming Press Books & Videos is a division of Miller Freeman Professional Ltd which provides a wide range of media services in agriculture and allied businesses. Among the magazines published by the group are:
Arable Farming, Dairy Farmer, Farming News, Pig Farming and **What's New in Farming.**
For a specimen copy of any of these please contact the address above.

Until then ...

A day will come when with a handshake the farmer will pay his tribute to the work the Land Girls did to keep things going. Until then, the vast mechanised army of farm workers is winning the battle against the blockade. And Ford and Fordson dealers are playing a big part by helping farmers to keep their tractors at work. The Ford repair and maintenance service cuts out delays and hold-ups on the farm—and saves fuel for the nation.

TRACTOR FUEL — WE WON'T WASTE IT. SAILOR

Farm by Ford or Fordson

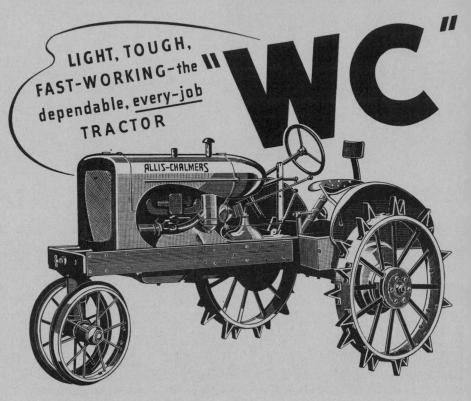

LIGHT, TOUGH, FAST-WORKING—the dependable, every-job TRACTOR "WC"

THE ALLIS-CHALMERS W.C. Model TRACTOR is strong enough to do the heaviest ploughing yet light enough to allow a girl to do a man-sized job on your farm every day. Shock-proof steering. Handy controls. Speedy and responsive engine gives untiring two-plough power with exceptional fuel economy. Allis-Chalmers —the Farmers' Priority.

ALLIS-CHALMERS
MANUFACTURING CO.

ABBEYDORE, HEREFORDSHIRE
Telephone : PONTRILAS 258-9 (2 lines)
Telegrams: "GYRATING ABBEYDORE"

TOTTON, SOUTHAMPTON
Telephone : 81461 and 81462